STUDENT NAME_____

GRADE_____

SCHOOL_____

TEACHER_____

BJC MATHEMATICS FORMAT

PAPER ONE		
DURATION	**WEIGHTING**	**STRUCTURE**
1 hour	40%	Short answer questions requiring little working. The content is taken from the entire syllabus.

PAPER TWO		
DURATION	**WEIGHTING**	**STRUCTURE**
2 hours	60%	More structured type questions of a more complex nature than those in paper one. The content is taken from the entire syllabus.

PRE-TEST

FOR OFFICIAL USE ONLY	
TOTAL MARKS	/50

SCHOOL NUMBER	CANDIDATE NUMBER
INITIAL(S)	SURNAME

INSTRUCTIONS

- Write your school number, candidate number as well as your initial(s) and surname in the spaces provided.
- Answer **ALL** questions in the spaces provided.
- **ALL** working <u>must</u> be shown,
- The use of calculators, slide riles, tables or other calculation aids is **NOT** allowed.
- **ALL** working is to be done in **blue** or **black ink**. Working and answers written in pencil, except **constructions and graphs**, may not be marked.
- **ALL** diagrams are not drawn to scale unless otherwise indicated.
- The mark for each question, or part question is shown in brackets [ ]

BJC Mathematics Practice Tests – Paper One

Answer **ALL** questions. Show **ALL** necessary working.

1. (a) 6054
     + 2137
       856
     ‾‾‾‾

   (b) 4937
       -2918
       ‾‾‾‾

Answer _____ [1]   Answer _____ [1]

(c) 5108
    X 7
   ‾‾‾‾

(d) 6084 ÷ 4

Answer _____ [1]   Answer _____ [1]

2. **January 2014**

Sun	Mon	Tues	Wed	Thur	Fri	Sat
			1	2	3	4
5	6	7	8	9	10	11
12	13	14	15	16	17	18
19	20	21	22	23	24	25
26	27	28	29	30	31	

Write down, as a fraction, " three days out of the total number of days in January'

Answer _____ [1]

BJC Mathematics Practice Tests – Paper One

3. Round 62.978 to the nearest whole number.

Answer _____ [1]

4. I am facing south west. I turn through an angle of 900 in a clockwise direction.

In which direction am I now facing?

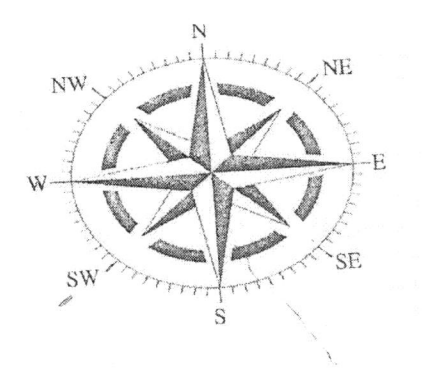

Answer _____ [1]

5. Write down the next two terms in the sequence.

16, 14, 12, 10, _____, _____ [2]

6. The clock shows 6:30 p.m. What time was it 4 hours and 30 minutes before 6:30 p.m.?

Answer _____[2]

7.

   (a)   What fraction of the diagram is NOT shaded?

Answer _____[1]

   (b)   Express the shaded part as a decimal.

Answer _____[1]

8. In prime factor form:

$$8 = 2 \times 2 \times 2$$

$$12 = 2 \times 2 \times 3$$

What is the LCM (Lowest Common Multiple) of 8 and 12?

Answer _____[2]

9. How much greater than 3.93 is 7.24?

Answer _____[2]

10. Fill in the square and cube of the given number.   [2]

Number	Number squared	Number Cubed
3		

11. Arrange these fractions in order from the least to the greatest.

$$\frac{1}{2}, \frac{5}{6}, \frac{2}{3}, \frac{1}{12},$$

Answer _____[4]

12. A circle has a diameter of 14 cm. Calculate the area of the circle ($\pi = \frac{22}{7}$)

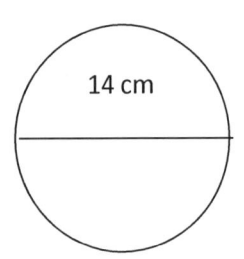

14 cm

Answer _____[3]

13. (a) Express $\frac{1}{5}$ as a percentage.

Answer _____[2]

(b) Calculate 15% of $200.

Answer _____[2]

14. Use the symbols <, > or = to make each statement true. [3]

(a) 5 cm ☐ 5 mm

(b) 500 kg ☐ 500 g

(c) 6l ☐ 6000 ml

BJC Mathematics Practice Tests –Paper One

15. Match the words in column 1 with the phrases in column 2 by putting the correct letter in the space provided. [5]

1	2
(a) Circumference	_____ plane figure with 6 sides
(b) Coefficient	_____ total distance around a circle.
(c) Hexagon	_____ angle greater than 180^0 but Less than 360^0.
(d) Reflex	_____ number before the variable.
(e) Quotient	_____ the result of division.

16. (a) Simplify

   $3x + 4y + 5x - 3y$

   Answer_____[2]

   (b) $C = \dfrac{88}{b+1}$ Calculate the value of C when b=10

   Answer_____[2]

17. (a) Calculate the size of angle **y**.

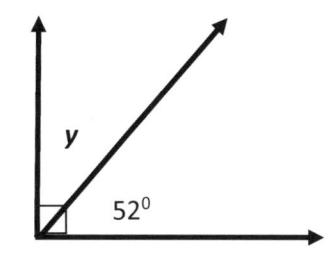

   52^0

   Answer_____[2]

BJC Mathematics Practice Tests – Paper One

(b) Calculate the size of angle w and angle x.

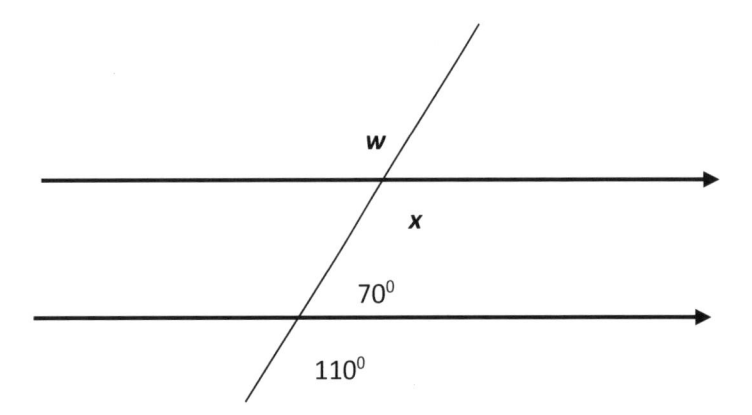

Answer w=_____[1]

Answer x =_____[2]

18.

James has x dollars. He lost $5.

(a) Write an algebraic expression for the amount he has left.

Answer =_____[1]

James now has $2 left.

(b) Write an algebraic equation to show all this information.

Answer =_____[1]

(c) Solve your equation formed in (b) to find 'x'.

Answer =_____[1]

PRACTICE TEST #1

FOR OFFICIAL USE ONLY	
TOTAL MARKS	/50

SCHOOL NUMBER	CANDIDATE NUMBER
INITIAL(S)	**SURNAME**

INSTRUCTIONS

- Write your school number, candidate number as well as your initial(s) and surname in the spaces provided.
- Answer **ALL** questions in the spaces provided.
- **ALL** working <u>must</u> be shown,
- The use of calculators, slide riles, tables or other calculation aids is **NOT** allowed.
- **ALL** working is to be done in **blue** or **black ink**. Working and answers written in pencil, except **constructions and graphs**, may not be marked.
- **ALL** diagrams are not drawn to scale unless otherwise indicated.
- The mark for each question, or part question is shown in brackets [ ]

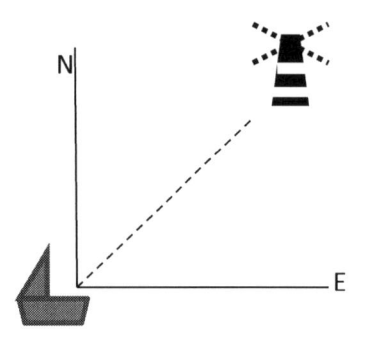

1) Using the diagram above, write down the three figure bearing of the lighthouse from the ship.

Answer_____[1]

2) (a)   348
        +567
        ─────
        ─────

   (b)   3489
        - 1625
        ─────

Answer (a)_____[1]

Answer (b)_____[1]

3) (a)   674
         x   5
         ─────

   (b)   1244÷4

                    Answer (a)_____[1]

                    Answer (b)_____[1]

4) Write down the next two prime numbers in the following sequence:
   11, 13, 17, 19, _____, _____

                    Answer _____[2]

5) Draw in the line(s) of symmetry on the figure below.

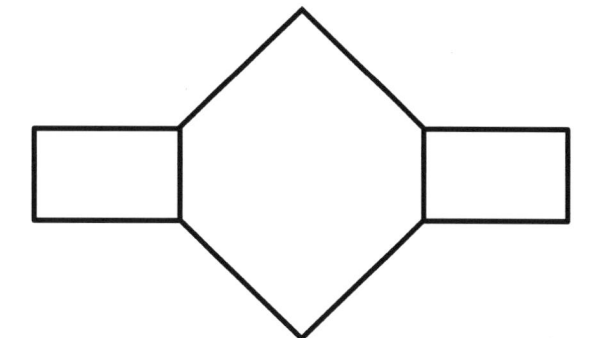

                              [2]

6) Write the ratio in its lowest terms.

    32 : 12

    Answer_____[2]

7) (a) Write 0.48 as a fraction.

    Answer_____[1]

(b) Write the answer in part (a) as a fraction in its lowest terms.

    Answer_____[1]

8) Given that **y=mx+c**, calculate the value of "**c**" when **y=15, m=3, and x=4**.

    Answer_____[3]

9) **Shade** the regions indicated, on the Venn diagrams below. [3]

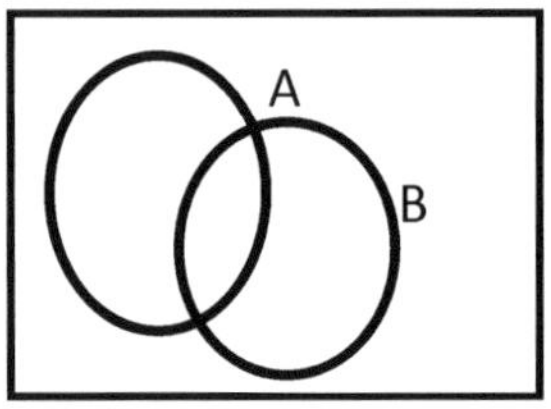

    $A \cup B$ 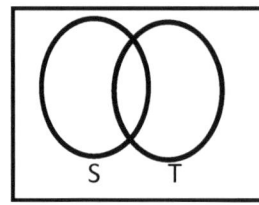 $S \cap T$ 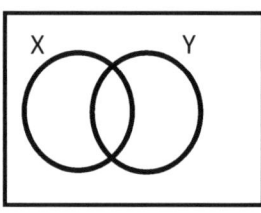 X'

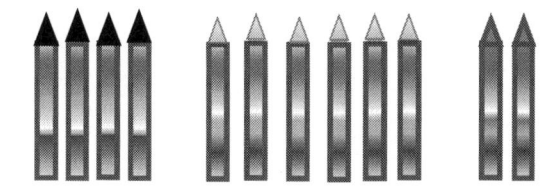

10) There are 4 black pens, 6 blue pens and 2 red pens in a box.
   (a) Find the probability of choosing, at random, a black pen.

   Answer_____[1]

   (b) Write the probability in (a) as a percent.

   Answer_____% [2]

11) Calculate the size of angle "m".

NOT TO SCALE

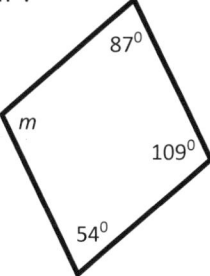

87°
m
109°
54°

Answer _____° [3]

12)

Kennae weighed 205 pound. She started a diet. The table below shows the results over the first three weeks

WEEKS	1	2	3
Weight gained	0 pounds	2 pounds	0 pounds
Weight lost	3 pounds	0 pounds	4 pounds

Calculate Kennae's weight at the end of the third week.

Answer _____ pounds [3]

13 (a) (i) Using a compass, ruler and pencil, **bisect** angle XYZ. [2]
   (ii) Name the bisector YM. [1]

X

Y   Z

(b) Measure and write down the size of angle MYZ.

Answer_____[1]

(a) From the set of numbers { 3, 4, 5, 6, 7 }, write down,
   (a) the even number(s),

Answer _____[2]

   (b) the multiple(s) of 3.

Answer _____[2]

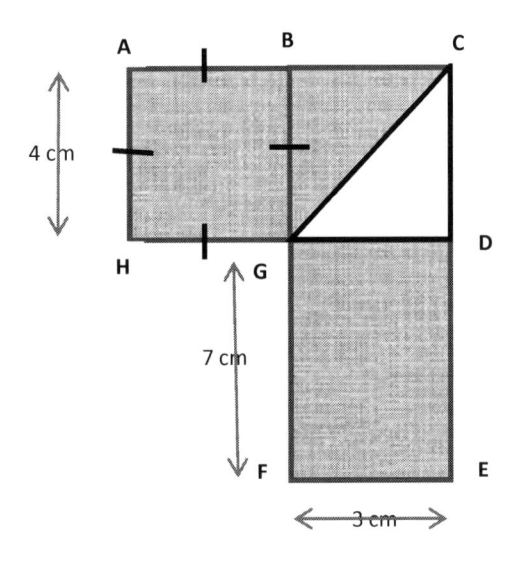

(b) Calculate the area of:
   (i) ABGH

Answer _____ cm² [1]

   (ii) BCG

Answer _____ cm² [1]

   (iii) GDEF

Answer _____ cm² [1]

   (iv) The total area of the shaded section.

Answer _____ cm² [1]

16. (a) Solve

$5p = 65$.

Answer _____ [2]

(b) Simplify,

(i) $5m + 3n + n - 2m$

Answer _____ [2]

(ii) $g^2 \times g^3$

Answer _____ [1]

17. Kamran played 7 games in a basketball tournament. His scores were:

30, 23, 16, 12, 20, 11 and 14.

Use the scores above to calculate,

(a) the **median** score,

Answer _____ [2]

(b) the **mean** (average) score.

Answer _____ [3]

PRACTICE TEST #2

FOR OFFICIAL USE ONLY	
TOTAL MARKS	/50

SCHOOL NUMBER	CANDIDATE NUMBER
INITIAL(S)	SURNAME

INSTRUCTIONS

- Write your school number, candidate number as well as your initial(s) and surname in the spaces provided.
- Answer **ALL** questions in the spaces provided.
- **ALL** working _must_ be shown,
- The use of calculators, slide riles, tables or other calculation aids is **NOT** allowed.
- **ALL** working is to be done in **blue** or **black ink**. Working and answers written in pencil, except **constructions and graphs**, may not be marked.
- **ALL** diagrams are not drawn to scale unless otherwise indicated.
- The mark for each question, or part question is shown in brackets [ ]

1. (a) 3020
       170
         5
     + 34
     ─────

Answer _____[1]

(b) 9875
    -1940
    ─────

Answer _____[1]

2. (a) 439
       x 5
       ────

Answer _____[1]

(b) 723÷9

Answer _____[1]

3. Write the ratio 36 : 56 in its lowest terms.

Answer _____[1]

4. From the diagram below, write down the **sum** of <a + <b + <c.

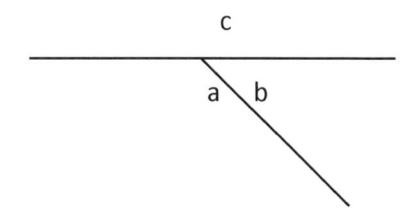

Answer _____ [1]

5. Circle the figure that represents $\frac{5}{6}$.

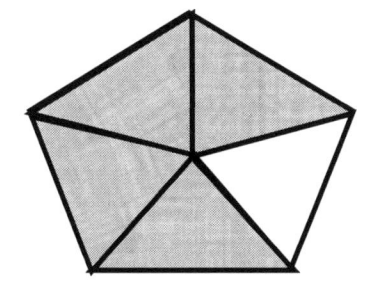

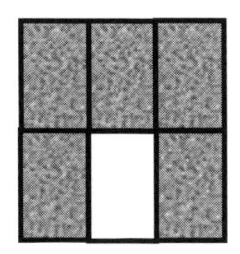

[1]

6. Use your protractor to draw an angle equal in size to the angle shown below. Use the line **AB**.

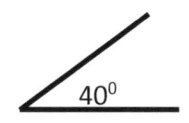

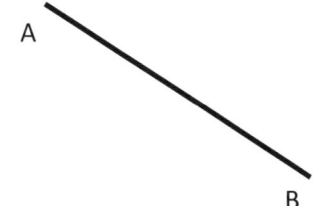

40°

[1]

BJC Mathematics Practice Tests –Paper One

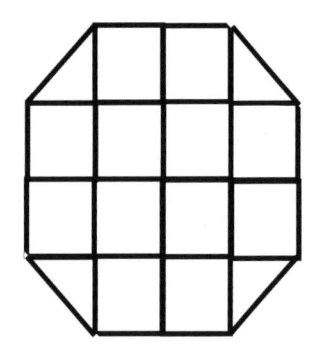

7. The area of each square is 1 square centimeter. Write down the area of the figure shown above.

Answer_____cm² [1]

8. Express 0.75 as a fraction in its lowest terms.

Answer_____ [2]

9. The pictures show High Tension Electrical Wires. Using the words listed below, name the types of lines suggested by each picture.

Parallel, intersecting, perpendicular

(a)

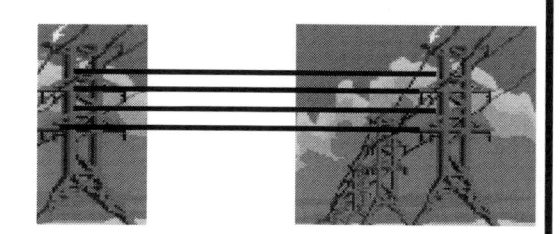

Answer_____[1]

(b)

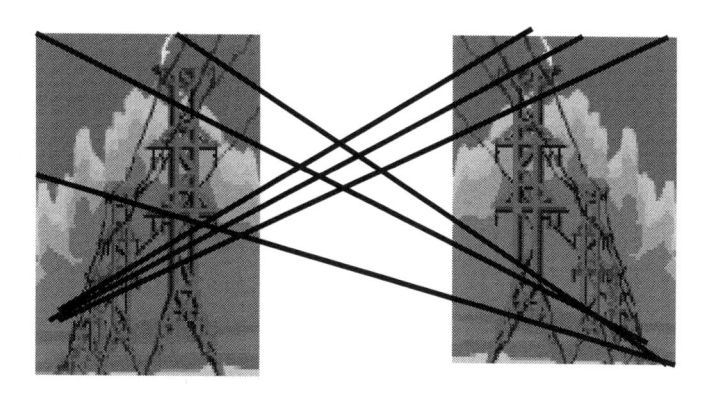

Answer_____[1]

10. A school's playground, measuring 33 meters, 43 meters, 50 meters and 25 meters, is enclosed by a chain link fence. How many meters of fencing is used.

Answer _____[2]

11. Draw in the line(s) of symmetry in the shape below.

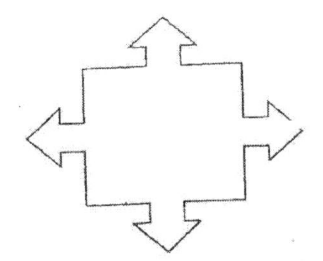

[2]

12. Use the list below to name the angle formed by the hands of each clock.

RIGHT ANGLE, ACUTE ANGLE, OBTUSE ANGLE

(a)

Answer _____ [1]

(b)

Answer _____ [1]

(c)

Answer _____ [1]

13. (a) Calculate the sum of 238 and 439.

Answer _____ [2]

(b) Write your answer to part (a) correct to the nearest ten.

Answer _____ [1]

14. Simplify
(a) $2g - 3g + 5g$

Answer _____ [1]

Solve

(b) $3a - 9 = 0$

Answer _____ [2]

15. Rewrite the following sentences, using set notation.
(a) 12 is a member of the set G.

Answer _____ [1]

(b) D is an empty set.

Answer _____ [1]

(c) The number of elements in set B is 10.

Answer _____ [1]

16. Given that a=-1, b=4, c= -3, find the value of

$$\frac{c+b}{b+a}$$

Answer _____ [3]

17. Calculate 60% of 25 meters.

Answer _____ m [3]

18. Mrs. Burrows earns $30,483 per year.(annually)
    (a) How much money does she earn in one month?

Answer $ _____ [3]

    (b) Write your answer in (a) correct to the nearest dollar.

Answer $ _____ [1]

19. Calculate the value of
    (0.81 ÷ 0.09) + (0.81 ÷ 9)

    Answer _____ [5]

20. (a) On the graph paper below, plot and join the following points in the order given.
    A = (2,2), B = (4,2), C = (6,4), D = (0,4). [5]

    (b) Give the special name of the quadrilateral formed.

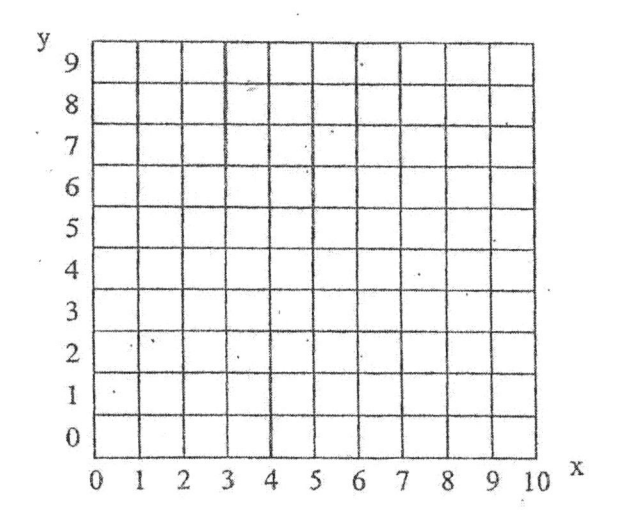

    Answer_____ [1]

PRACTICE TEST #3

FOR OFFICIAL USE ONLY	
TOTAL MARKS	/50

SCHOOL NUMBER	CANDIDATE NUMBER
INITIAL(S)	SURNAME

INSTRUCTIONS

- Write your school number, candidate number as well as your initial(s) and surname in the spaces provided.
- Answer **ALL** questions in the spaces provided.
- **ALL** working *must* be shown,
- The use of calculators, slide riles, tables or other calculation aids is **NOT** allowed.
- **ALL** working is to be done in **blue** or **black ink**. Working and answers written in pencil, except **constructions and graphs**, may not be marked.
- **ALL** diagrams are not drawn to scale unless otherwise indicated.
- The mark for each question, or part question is shown in brackets [ ]

Answer ALL questions. Show ALL necessary working.

1. (a) 354
       + 35
       649
       ____

       Answer_____[1]

   (b) 7858
       - 5287
       ____

       Answer_____[1]

2. (a) 2875
       x   3
       ____

       Answer _____[1]

   (b) 4896÷4

       Answer_____[1]

3. Draw in the line(s) of symmetry on the shape below.

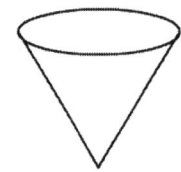

   [1]

4. Write the decimal that represents the shaded portion of the figure below.

Answer_____[2]

5. When a = 5, b = 7 and c = 0, calculate the value of:

3a + b + c

Answer_____[2]

6. Write the decimal 0.65 as:

(a) a fraction in its lowest terms.

Answer_____[2]

(b) a percentage.

Answer_____[2]

7. Triangle ABC is given below.

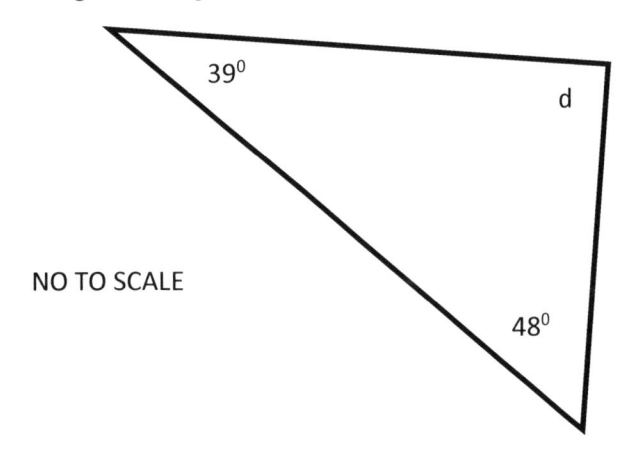

    39°

    d

    NO TO SCALE

    48°

    (a) Calculate the size of angle d.

    Answer _____ [2]

    (b) What special type of triangle is ABC?

    Answer _____ [1]

8. One pumpkin weighs 1.2 kg. Another pumpkin weighs 108.6 g. What is the total weight, in grams, of both pumpkins?

    Answer _____ g [2]

9.

Wow! how many out of 12? What if I had made 100% of the shots?

Samuel plays on the school's basketball team. In one game he shoots 12 free throws and makes 75% of them. How many free throws did he make?

Answer_____[2]

10. (a) Solve for a.

$a + 6 = 24$

Answer_____[2]

(b) Simplify:
(i) $h^5 \div h^2$

Answer_____[1]

(ii) $6z - 4z + 8z$

Answer_____[1]

11. (a) List ALL of the prime numbers found on the stars below.

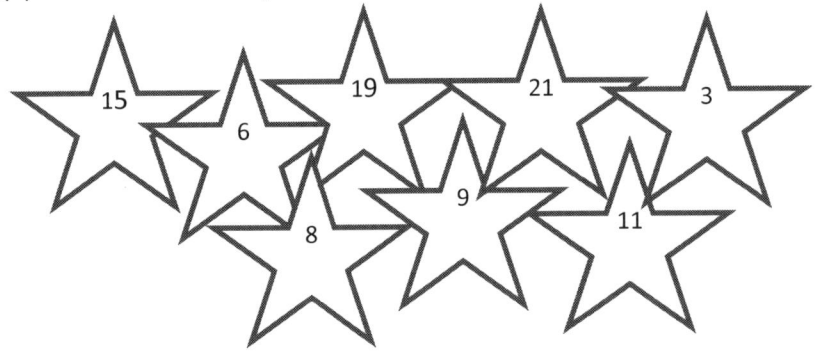

[Stars containing: 15, 19, 21, 3, 6, 8, 9, 11]

Answer_____[2]

(b) Calculate the sum of the prime numbers in (a).

Answer_____[2]

12. There are 15 girls and 12 boys in a class.

(a) Express the number of boys to the total number of students in the class as a ratio.

Answer_____[2]

(b) Write your answer to (a) in its lowest terms.

Answer_____[2]

5 pens cost $7.45. Calculate the cost of:

   (a)   one pen,

   Answer _____[2]

   (b)   3 pens.

   Answer_____[2]

14.

TICKETS FOR CARNIVAL RIDES	
Ferris Wheels	6 tickets
Carousel	3 tickets
Go Car	2 tickets

   (a)   How many tickets do I need if I want to go on each ride once?

   Answer_____ tickets [1]

   (b)   Each ticket costs 25 cents. How much money would I spend for tickets to go on each ride once?

   Answer_____[2]

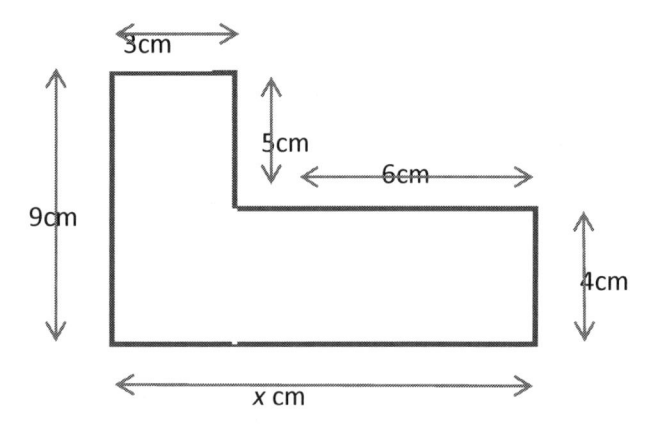

(a) Write down the length of x.

Answer _____ cm [2]

(b) Calculate the perimeter of the above figure.

Answer _____ cm [2]

16. Study the calendar below, then answer the questions which follow.

November 2000

S	M	T	W	T	F	S
			1	2	3	4
5	6	7	8	9	10	11
12	13	14	15	16	17	18
19	20	21	22	23	24	25
26	27	28	29	30		

(a) Bruce's birthday is on the first Friday in November. What date would that be?

Answer_____[1]

(b) What day of the week is two weeks after the 15th?

Answer_____[1]

(c) The Sunday nearest to November 10th is Remembrance Sunday. What date was Remembrance Sunday in 2000?

Answer_____[1]

17. Jodenell is training for a marathon. He runs $1\frac{1}{8}$ miles on Sunday, $2\frac{1}{3}$ miles on Monday and $1\frac{1}{4}$ miles on Tuesday. How far does he run altogether?

Answer_____ miles [4]

PRACTICE TEST #4

FOR OFFICIAL USE ONLY	
TOTAL MARKS	/50

SCHOOL NUMBER	CANDIDATE NUMBER
INITIAL(S)	SURNAME

INSTRUCTIONS

- Write your school number, candidate number as well as your initial(s) and surname in the spaces provided.
- Answer **ALL** questions in the spaces provided.
- **ALL** working *must* be shown,
- The use of calculators, slide riles, tables or other calculation aids is **NOT** allowed.
- **ALL** working is to be done in **blue** or **black ink**. Working and answers written in pencil, except **constructions and graphs**, may not be marked.
- **ALL** diagrams are not drawn to scale unless otherwise indicated.
- The mark for each question, or part question is shown in brackets [ ]

1.
(a)   204
     + 93
      581
     _____

Answer _____[1]

(b)   7483
    - 5643
    _____

Answer _____[1]

(c)   734
     x   3
     _____

Answer _____[1]

(d)   8094 ÷ 6

Answer _____[1]

2. Write down the number that has

4 in the hundreds place, **7** in the tens place, **0** in the thousands place, **8** in the units(ones) and **2** in the ten thousands place.

Answer _____[1]

BJC Mathematics Practice Tests – Paper One   *Page 42*

3. Fill in the blank in the sequence below.

1, 3, 4, 7, 11, 18 _____

Answer _____[1]

4.

NAME: Alanna Brown

MATHEMATICS TEST

$\dfrac{65}{100}$

Alanna scored 65 out of 100 on Mrs. Burrows' Mathematics test. Express this score as a percent.

Answer _____[1]

5. Write down all the prime numbers between 60 and 70.

Answer _____[2]

6. Choose the correct symbol from the box below that will make each statement true.

$$\notin, \subset$$

(i) {April} _____ { month of the year}

(ii) 15 _____ { multiples of 5}

Answer _____[1]
Answer _____[1]

BJC Mathematics Practice Tests – Paper One

7. Express 0.4 as a fraction in its lowest terms.

Answer _____ [2]

8. The prime factors of two numbers are listed below.

$$2 \times 2 \times 3 \times 3 \times 5$$
$$2 \times 2 \times 2 \times 3 \times 5$$

Write down the Highest/Greatest Common Factor (HCF/GCF) of these numbers.

Answer _____ [2]

9. The spinner shows four colors.

RED	GREEN
BLUE	PINK

What is the probability of getting a pink on the first spin?

Answer _____ [2]

10.

Sawyer's Peas
0.45 kg

A tin of peas weighs 0.45 kg. Calculate the total weight of a dozen such tins.

Answer _____ [3]

11.

Island airways leaves Nassau at 8:10 am and arrives in Inagua at 10:05 am How long does the flight last?

Answer _____ [3]

12. Using the diagram and the list below, choose the name that best describes the labeled shapes.

| Parallelogram | Trapezium | Right-angled triangle |

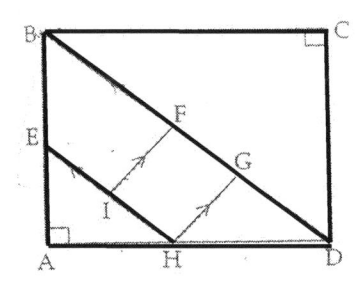

(a) BEIF

Answer _____[1]

(b) BCD

Answer _____[1]

(c) FGHI

Answer _____[1]

(d) Name the diagonal

Answer _____[1]

13. (a) Write down the size of angle PQR shown in the diagram below.

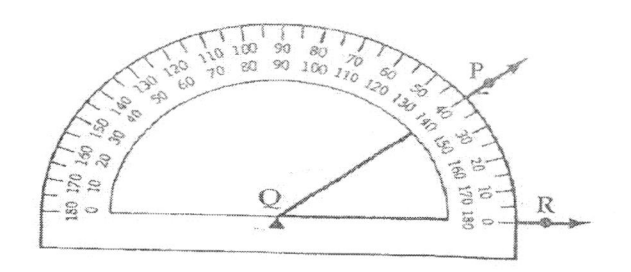

Answer _____ [1]

(b) 3. Identify each angle formed by the letters *a,b,c* and *d.*

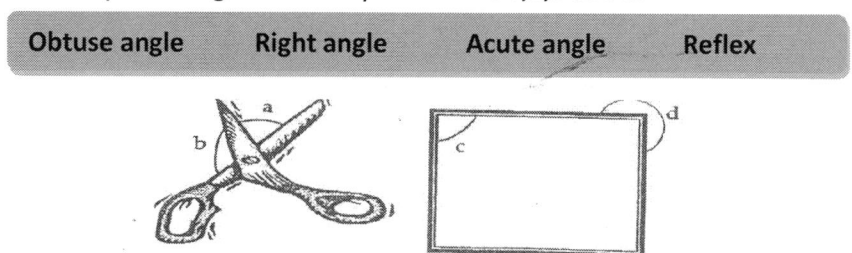

| Obtuse angle | Right angle | Acute angle | Reflex |

Answer a _____ [1]

Answer b _____ [1]

Answer c _____ [1]

Answer d _____ [1]

BJC Mathematics Practice Tests –Paper One

| 4.6 kg | 2kg | 2.8kg | 3.6kg |

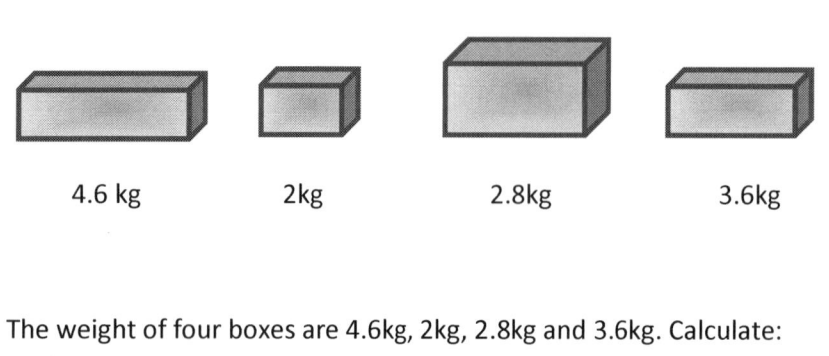

The weight of four boxes are 4.6kg, 2kg, 2.8kg and 3.6kg. Calculate:

(a) the **total** weight of the boxes.

Answer _____[1]

(a) the **average** weight of the boxes.

Answer _____[2]

(c) Express the weight of the lightest box in grams.

Answer _____[2]

15. (a) Shade $\frac{3}{8}$ of the figure below. [1]

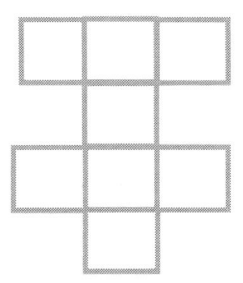

(b) Which fraction is larger $\frac{7}{10}$ or $\frac{4}{5}$?

Answer _____ [2]

(b) What is the **difference** between the larger and smaller fraction, $\frac{7}{10}$ and $\frac{4}{5}$?

Answer _____ [3]

BJC Mathematics Practice Tests –Paper One *Page 49*

16. Simplify

(a) 3b + 5b − 2b

Answer _____[1]

(b) (i) Solve

y + 3 = 15

Answer _____[2]

(ii) 9h = 108

Answer _____[2]

(iii) $\frac{x}{2} = 12$

Answer _____[2]

PRACTICE TEST #5

FOR OFFICIAL USE ONLY	
TOTAL MARKS	/50

SCHOOL NUMBER	CANDIDATE NUMBER
INITIAL(S)	SURNAME

INSTRUCTIONS

- Write your school number, candidate number as well as your initial(s) and surname in the spaces provided.
- Answer **ALL** questions in the spaces provided.
- **ALL** working *must* be shown,
- The use of calculators, slide riles, tables or other calculation aids is **NOT** allowed.
- **ALL** working is to be done in **blue** or **black ink**. Working and answers written in pencil, except **constructions and graphs**, may not be marked.
- **ALL** diagrams are not drawn to scale unless otherwise indicated.
- The mark for each question, or part question is shown in brackets [ ]

1.
(a)   6035
      + 721
         89
          4
      _____

Answer _____ [1]

(b)   8095
      - 3726
      _____

Answer _____ [1]

(c)   745
      x   6
      _____

Answer _____ [1]

(d)   6321 ÷ 7

Answer _____ [1]

2. Shade $\frac{2}{3}$ of the triangles below.

△ △ △ △ △ △

[1]

3. Measure and write down the length of AB in centimetres.

A ——————————————— B

Answer _____ [1]

4. (a) Fill in the box to create an improper fraction. [1]

$$\frac{\Box}{7}$$

(b) Express your answer to (a) as a mixed fraction.

Answer _____ [1]

5. If $x = 9$, evaluate

$x - \sqrt{25}$

Answer _____ [2]

6. Use >, < or = to make each statement true.

(a) 25 + 25 ☐ 40   [1]

(b) 6.03 ☐ 6.3   [1]

7. Write the next two terms in the following sequence

1, 3, 6, 10, _____, _____   [2]

8. The sum of three numbers is 156. Two of the numbers are 63 and 49. What is the third number?

Answer =_____   [2]

9. In a test, Lyndeisha did the following sum:

$$\begin{array}{r} 6.7 \\ \times\, 0.41 \\ \hline 67 \\ 26\,8 \\ \hline 27.47 \end{array}$$

(a)   Write down Lyndeisha's mistake.

       Answer _____   [1]

(b)   Give the correct answer to this sum.

       Answer: _____   [1]

BJC Mathematics Practice Tests –Paper One

10.

(a) Write down the temperature shown at A.

Answer: _____ [1]

(b) Mark and label the spot **Y** which is 15° below 0.   [1]

11. Write down the lengths in order of size, starting with the shortest.

6 cm, 16 mm, 2 m, 4 dm

Answer _____ [2]

12.

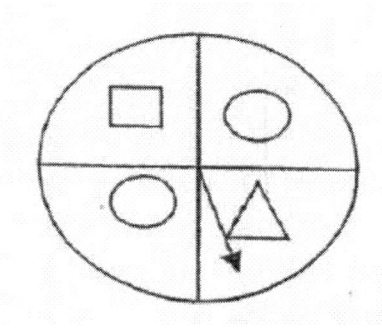

The diagram shows a spinner. The spinner is spun. What is the probability of the pointer landing on

(a) a triangle?

Answer _____[1]

(b) a pentagon

Answer _____[1]

13. (a) Write down the perimeter of the shape below.

Answer _____[1]

(b) If a = 4cm, b = 3 cm and c = 3.5 cm, calculate the perimeter of the shape.

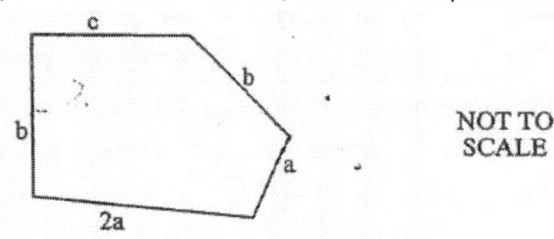

NOT TO SCALE

Answer _____[2]

14. Solve:

(a) 4x = 20

Answer _____ [1]

(b) 2m + 3 = 9

Answer _____ [2]

15. (a) Calculate the value of ∠a.

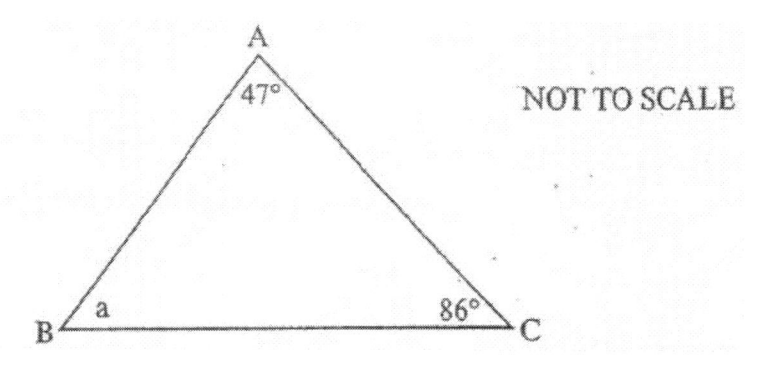

NOT TO SCALE

Answer: _____ ° [2]

(b) Name the two equal sides in this triangle.

Answer _____ [1]

16. Use the list of words given to identify the geometrical shapes shown.

cone, cube, cuboid, cylinder, sphere

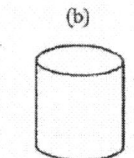

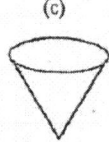

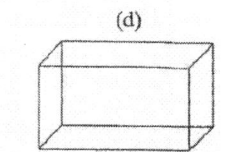

(a)   (b)   (c)   (d)

Answers: (a) _____ [1]

(b) _____ [1]

(c) _____ [1]

(d) _____ [1]

17. The banner for AF school requires $3\frac{1}{2}$ yards of blue ribbon, $2\frac{1}{3}$ yards of white ribbon and $4\frac{1}{6}$ of aqua ribbon. Calculate the total length of ribbon needed.

Answer _____ yards [4]

18.

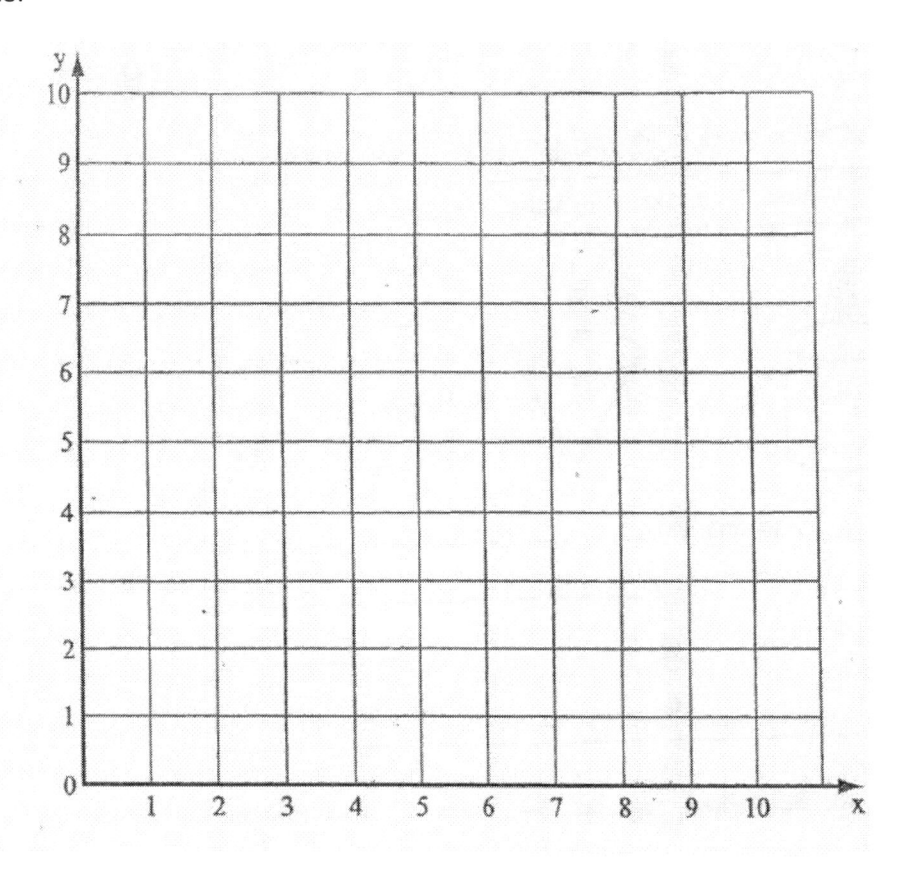

(a) Use the graph above to plot and join the following points in the order in which they are written:

A(3,7)→B(5,7) →C(6,5) →D(6,3) →E(5,1) →F(3,1) →G(2,3) →H(2,5).
Now join H to A.

[3]

(b) Write down the special geometrical given to the figure formed.

Answer: _____ [1]

19.

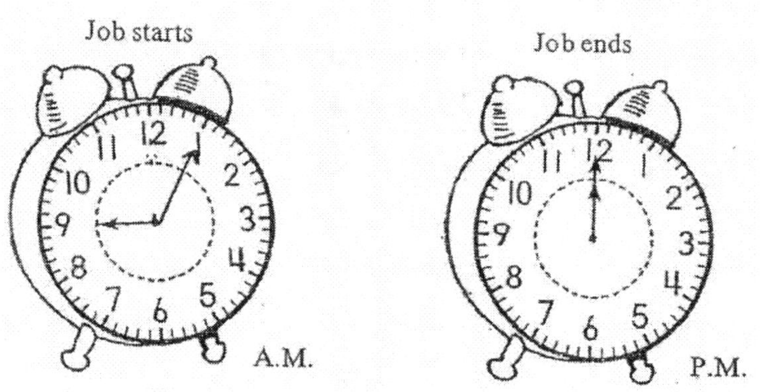

(a) At what time did the job start?

Answer: ______________________________ [1]

(b) What time did the job end?

Answer: ______________________________ [1]

(c) How long did the job last?

Answer: ______________________________ [1]

(d) Lunch on board the Sea Link is served at 1400 hours. Express this time using the 12 hour clock.

Answer:______________________________ [2]

PRACTICE TEST #6

FOR OFFICIAL USE ONLY	
TOTAL MARKS	/50

SCHOOL NUMBER	CANDIDATE NUMBER
INITIAL(S)	SURNAME

INSTRUCTIONS

- Write your school number, candidate number as well as your initial(s) and surname in the spaces provided.
- Answer **ALL** questions in the spaces provided.
- **ALL** working <u>must</u> be shown,
- The use of calculators, slide riles, tables or other calculation aids is **NOT** allowed.
- **ALL** working is to be done in **blue** or **black ink**. Working and answers written in pencil, except **constructions and graphs**, may not be marked.
- **ALL** diagrams are not drawn to scale unless otherwise indicated.
- The mark for each question, or part question is shown in brackets [ ]

1.
(a) 3024
    + 853
     9671
     ─────

Answer _____[1]

(b)   8367
    - 5942
     ─────

Answer _____[1]

2.
(a)   856
    x   5
     ─────

Answer _____[1]

(b)   9036 ÷ 4

Answer _____[1]

3. Yolanda and Rolanda were given a project to do together. Yolanda completed $\frac{3}{8}$ of it and Rolanda completed $\frac{1}{4}$ of it. How much of the project was completed?

Answer _____[3]

4. A blackout began at 5:20 p.m and ended at 7:05 p.m. How long did the blackout last?

Answer _____hr(s) _____mins   `[2]

5.   NOT TO SCALE

A   B   C   E   M   F   G   H   I

     1 inch   2 inches   3 inches   4 inches

a. How many units does the line segment BF represent?

Answer _____[1]

b. Which letter represents $2\frac{1}{2}$ units from A?

Answer _____[1]

BJC Mathematics Practice Tests – Paper One   *Page 63*

6. Use a symbol from the box that will make each statement true.

$$> \quad < \quad =$$

a. $\frac{1}{4}$ ☐ 0.36 [1]

b. 75% ☐ 0.75 [1]

c. $\frac{3}{5}$ ☐ 25% [1]

7. a. Evaluate $3^3 + 1$

Answer _____[2]

b. State whether your answer to (a) is a Prime number or a Composite number.

Answer _____[1]

8. a. Calculate the sum of: 0.1, 3, 6.5 and 0.75.

Answer _____[2]

b. Round your answer in (a), to the nearest whole number.

Answer _____[1]

BJC Mathematics Practice Tests –Paper One

9.

Days of the week	Monday	Tuesday	Wednesday
Temperature	-2°	-10°	5°

The temperatures were recorded for the three days in New York city.

a. On which day was the temperature the lowest?

Answer _____ [1]

b. Name the day on which the temperature was the highest.

Answer _____ [1]

c. What is the difference between the temperature on Wednesday and the temperature on Tuesday?

Answer _____ °[2]

10. Write the next two terms in each of the following sequences:

a. 3, 9, 15, 21, _____, _____ [2]

b. 29, 31, 37, 41, _____, _____ [2]

11. Use the best word from the list to name each polygon below.

| Rhombus   Trapezium   Hexagon   Octagon |

(a) [hexagon shape]

(b) [octagon shape]

(c) [trapezium shape]

(d) [rhombus shape]

Answer (a) _____ [1]

Answer (b) _____ [1]

Answer (c) _____ [1]

Answer (d) _____ [1]

12. Use a pencil and a pair of compasses to answer this question. Show ALL construction lines clearly.

(a) Construct an angle of $60°$ at A.

A _____

[3]

(b) Name the angle in (a) < XAM   [1]

BJC Mathematics Practice Tests –Paper One

13. Evaluate.

(a) 5 + 7 x 2 - 18

Answer _____[2]

(b) 24 ÷ 4 + 24 ÷ 6

Answer _____[2]

14. Solve for a.

$$6a = 33$$

Answer _____[2]

(b) When d=2, and c=4, find the value of

$$\frac{8d + c}{c}$$

Answer _____[2]

15.

Terwashaan's Fashions

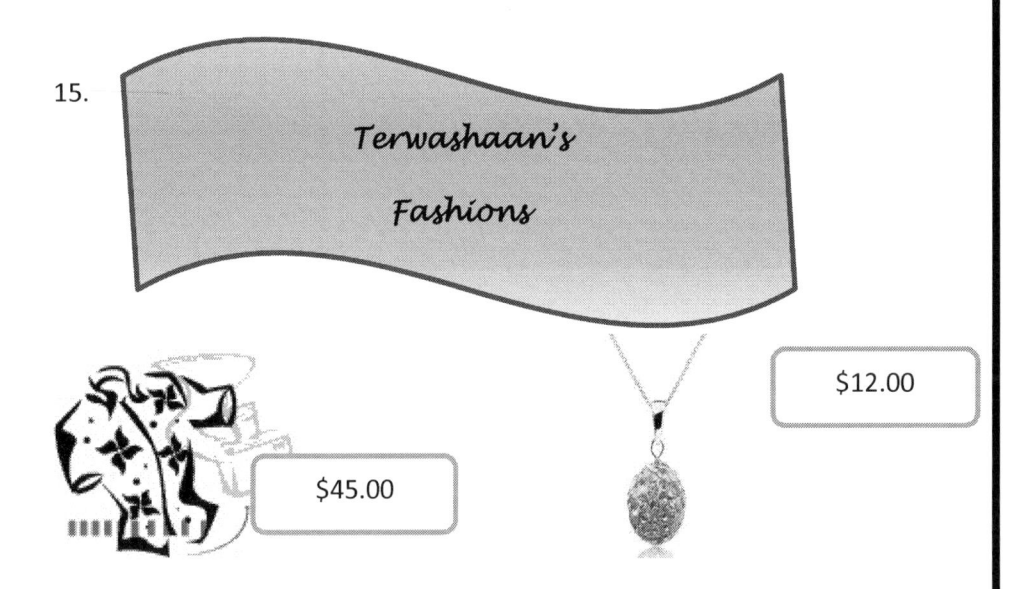

$45.00

$12.00

Miracle bought a blouse for $45 and a necklace for $12 at Terwashaan's Fashions.

(a) How much did she have to pay?

Answer _____[1]

Melissa Sears Fashions

$47.00

$13.00

Lashan bought the same items from Melissa Sears Fashions.

(b) How much is her total?

Answer _____[1]

Melisssa Sears gives Lashan 10% off.

(c) How much does Lashan actually pay for the items?

Answer _____[3]

(d) Who gets the better deal, Miracle or Lashan?

Answer _____[1]

PRACTICE TEST #7

FOR OFFICIAL USE ONLY	
TOTAL MARKS	/50

SCHOOL NUMBER	CANDIDATE NUMBER
INITIAL(S)	SURNAME

INSTRUCTIONS

- Write your school number, candidate number as well as your initial(s) and surname in the spaces provided.
- Answer **ALL** questions in the spaces provided.
- **ALL** working *must* be shown,
- The use of calculators, slide riles, tables or other calculation aids is **NOT** allowed.
- **ALL** working is to be done in **blue** or **black ink**. Working and answers written in pencil, except **constructions and graphs**, may not be marked.
- **ALL** diagrams are not drawn to scale unless otherwise indicated.
- The mark for each question, or part question is shown in brackets [ ]

1.
(a)   457
      + 150
        231
      _____

        Answer _____[1]

(b)   8654
     - 1832
      _____

        Answer _____[1]

2.
(a)   1104
     x   3
      _____

        Answer _____[1]

(b)   828 ÷ 4

        Answer _____[1]

3. Mrs. Miller's children drink 1 ½ gallons of milk each week. How many
   Gallons of mild will they drink in six weeks?

        Answer _____[2]

BJC Mathematics Practice Tests –Paper One   *Page 72*

4. Calculate the size of angle **P**.

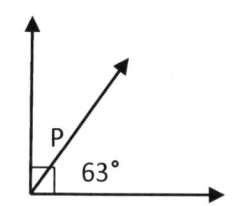

NOT TO SCALE

Answer _____ [2]

5. Using PENCIL, COMPASS AND RULER ONLY, BISECT PQR

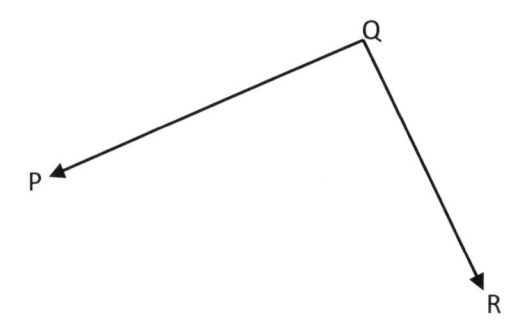

[3]

6. Insert the correct symbols from the list into each box to make each statement true.

$$+, -, \times, \div$$

(a) ¾ ☐ ¼ = 1   [1]

(b) ¾ ☐ ¼ = 3   [1]

(c) ¾ ☐ ¼ = 3/16   [1]

(d) ¾ ☐ ¼ = ½   [1]

7. Lakecia bought one of each of the following items:

   Iron - $17.99
   Microwave - $69.00

   (a) Calculate her total bill.

   Answer _____ [1]

   Lakecia paid her bill with two $50 notes.

   (b) How much change should she receive?

   Answer _____ [3]

8.

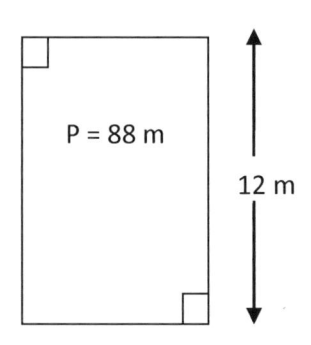

P = 88 m

12 m

The diagram represents a garden. The perimeter (P) of the garden is 88 metres. The length of the garden is 12 metres. Calculate the width (breadth) of the garden.

Answer _____[4]

9.

Yellow Elder PUBLIC SCHOOL

Yellow Elder Public School had an enrollment of 1200 students at the beginning of the new school year. There was a 20% increase in the enrollment at mid term.

(a) By how many students was the enrolment increased?

Answer _____[2]

(b) What is the new enrolment?

Answer _____[2]

10.(a)　(i) Express 60% as a fraction.

Answer _____ [1]

(ii) Write your answer in (a)(i) in its lowest terms.

Answer _____ [1]

(b) Express ¾ as a decimal.

Answer _____ [2]

11. Five pens cost $7.45. Find the cost of:

(a) One pen,

Answer _____ [2]

(b) Fifteen such pens,

Answer _____ [2]

12. There are four oranges, two apples and three bananas in a fruit basket.

(a) Write as a ratio, in its lowest terms, the number of apples to the number of oranges.

Answer _____ [2]

(b) What is the probability of picking a banana out of the fruit basket?

Answer _____[2]

13.

BLUE HILL ROAD HIGH SCHOOL				
Name: Jamilia Carroll **Grade:** 8 **Merits:** 146			**Age:** 13 years **Absent:** 3 days **Demerits:** 0	
SUBJECT	**GRADE**	**EFFORT**	**CONDUCT**	**SIGNATURE**
English Language	64			
Mathematics	76			
Spanish	70			
General Science	60			
Physical Education	90			
Religious Studies	54			
Social Studies	76			

The report reflects Jamilia's grades at the end of the 2016-2017 school year. What is the

(a) Modal grade?

Answer _____[1]

(b) Median grade?

Answer _____[2]

(c) Mean grade?

Answer _____[2]

14. Simplify:

   (a) $12m + 4g - 8m + g$

   Answer _____ [2]

   (b) When a = 4, find the value of

   $6a - 5$

   Answer _____ [2]

   (c) Solve for 'x'

   $X - 13 = 43$

   Answer _____ [2]

PRACTICE TEST #8

FOR OFFICIAL USE ONLY	
TOTAL MARKS	/50

SCHOOL NUMBER	CANDIDATE NUMBER
INITIAL(S)	SURNAME

INSTRUCTIONS

- Write your school number, candidate number as well as your initial(s) and surname in the spaces provided.
- Answer **ALL** questions in the spaces provided.
- **ALL** working *must* be shown,
- The use of calculators, slide riles, tables or other calculation aids is **NOT** allowed.
- **ALL** working is to be done in **blue** or **black ink**. Working and answers written in pencil, except **constructions and graphs**, may not be marked.
- **ALL** diagrams are not drawn to scale unless otherwise indicated.
- The mark for each question, or part question is shown in brackets [ ]

1.
(a) 248
    +3527
     603
    _____

        Answer _____[1]

(b) 7609
    - 4295
    _____

        Answer _____[1]

(c) 5841
    x   8
    _____

        Answer _____[1]

(d) 931 ÷ 7

        Answer _____[1]

2. Write the letter which indicates the reflex angle.

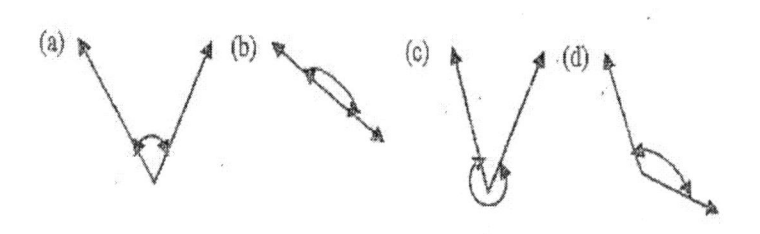

(a)   (b)   (c)   (d)

        Answer _____[1]

BJC Mathematics Practice Tests –Paper One

3. Write the number that is **NOT** a square number.

     49   **16**   **30**   **9**

     Answer _____[1]

4. Write the number that is **NOT** a prime number.

     2   **3**   **7**   **21**

     Answer _____[1]

5. Draw in the line of symmetry in the picture below.

[1]

6. Study the diagram below then answer the questions.

Write down
(a) One pair of vertically opposite angles.

     Answer _____[1]

(b) One pair of supplementary angles

     Answer _____[1]

7. I begin at due north and travel in a clock-wise direction to due west.
   Through how many degrees have I travelled?

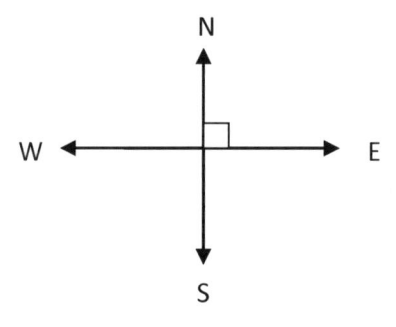

```
        N
        ↑
   W ←——⌐——→ E
        ↓
        S
```

Answer _____[1]

8. Find the value of

   $2^3 \times 3^2$

   Answer _____[3]

9.

24	42	51
6	89	27
49	31	65

From the numbers above, write down.

(a) a multiple of 4

Answer _____[1]

(b) a factor of 12

Answer _____[1]

(c) a cube number

Answer _____[1]

10. One book costs $8. How many books can I buy with $42.00?

Answer _____[2]

BJC Mathematics Practice Tests –Paper One

11. Calculate the size of angle M.

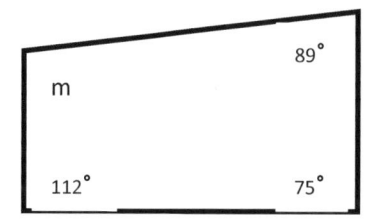

   (quadrilateral with angles 89°, 75°, 112°, and m)

   Answer _____ [3]

12. Evaluate

   $$\frac{3}{5} \times \frac{1}{4} \div \frac{3}{8}$$

   Answer _____ [3]

13. Use a pencil, a ruler and a pair of compasses to bisect the line TR.

T _____ R

[3]

14. From the list below, choose the name that describes the solids below.

| Cone | Cube | Cylinder | Pyramid |

(a)   (b)   (c)   (d)

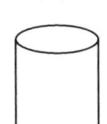

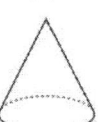

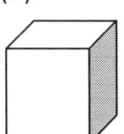

Answer (a) _____ [1]

(b) _____ [1]

(c) _____ [1]

(d) _____ [1]

5. | F | R | A | C | T | I | O | N | S |

A card is selected at random from the cards shown above. What is the Probability that it is

(a) A vowel? (a, i, o)

Answer _____ [2]

(b) One of the first three letters of the alphabet?

Answer _____ [2]

16.

Grade	Number of Students
A	3
B	10
C	14
D	3
E	5

The table represents the grades of a class in a recent test.

(a) How many students are in the class?

Answer _____ [1]

(b) What is the modal **grade**?

Answer _____ [1]

(c) To the nearest whole number, what percentage of the class received an 'A' grade?

Answer _____ [3]

17. The bar chart below represents the sport played by a group of 100 students.

SPORTS PLAYED

Spo rt

| Baseball |
| Football |
| Basketball |
| Tennis |

0   10   20   30   40   50
Number of students

(a) Draw in the bar to represent the number who played football.

[2]

(b) Which sport had the largest number of players?

Answer _____ [1]

(c) Which sport had ten (10) players more than tennis?

Answer _____ [1]

18. (a) Simplify by collecting the like terms.

   $8d - 5d + 3d$

   Answer _____ [1]

   (b) Solve for y.

   $y + 12 = 27$

   Answer _____ [2]

   (c) When $a = 3$, $b = 0$ and $c = 11$, find the value of

   $ab + c$

   Answer _____ [2]

PRACTICE TEST #9

FOR OFFICIAL USE ONLY	
TOTAL MARKS	/50

SCHOOL NUMBER	CANDIDATE NUMBER
INITIAL(S)	SURNAME

INSTRUCTIONS

- Write your school number, candidate number as well as your initial(s) and surname in the spaces provided.
- Answer **ALL** questions in the spaces provided.
- **ALL** working <u>must</u> be shown,
- The use of calculators, slide riles, tables or other calculation aids is **NOT** allowed.
- **ALL** working is to be done in **blue** or **black ink**. Working and answers written in pencil, except **constructions and graphs**, may not be marked.
- **ALL** diagrams are not drawn to scale unless otherwise indicated.
- The mark for each question, or part question is shown in brackets [ ]

1.
(a) 137
    + 62
     258
    ____

        Answer _____[1]

(b) 937
    - 897
    ____

        Answer _____[1]

2.
(a) 1208
    x   6
    ____

        Answer _____[1]

(b) 1796 ÷ 4

        Answer _____[1]

3. Write the letter of the diagram which shows a trapezium.

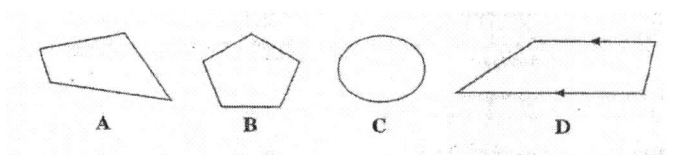

        Answer _____[1]

BJC Mathematics Practice Tests –Paper One   Page 90

4. Write the fraction represented by the shaded regions.

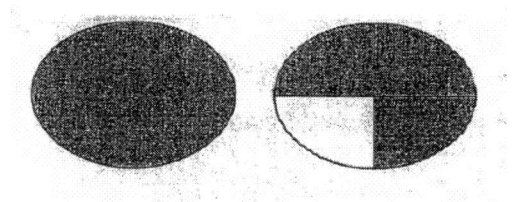

Answer _____[2]

5. Draw in the line(s) of symmetry on the diagram below.

[2]

6. On a map one inch represents five miles. How many inches would represent 35 miles?

Answer _____ inches [2]

7. When c = 3 and d = 4, evaluate $\frac{cd}{2}$

Answer _____[2]

8. Quincy bought a pair of shoes in a sale for $45.90. He saved $5.45. What was the regular price of the shoes?

Answer _____[2]

9. Study the diagram below then answer the questions which follow.

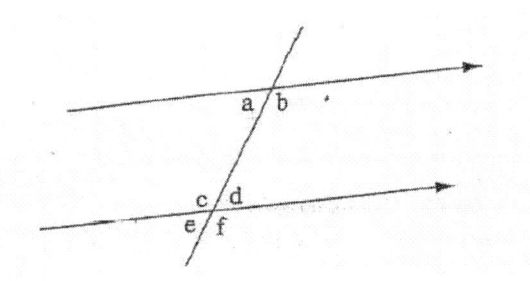

(a) Name **ONE** pair of corresponding angles.

Answer _____[1]

(b) Angle a = 60°, what is the size of angle d?

Answer _____[1]

10. Express 38% as a fraction in its lowest terms.

Answer _____ [2]

11. From the set of numbers {6,7,11,61,51}, write down:

(a) **ONE** composite number,

Answer _____ [1]

(b) **ONE** odd number,

Answer _____ [1]

(c) **ONE** factor of 54

Answer _____ [1]

12. The sum of two numbers is 12. One of the numbers is 8.78. Find the other number.

Answer _____ [3]

13. Evaluate:

$5^2 + 3^3$

Answer _____ [3]

14. (a) Simplify by collecting like terms.

   $16m - 4m + m$

   Answer _____[1]

   (b) Solve for *d*.

   $5d - 8 = 32$

   Answer _____[3]

15. Tickets marked **R, Y, B, W** were placed on a tray.

R	R	R		W
	R	W	Y	
Y		W	B	Y

   Without looking, what is the probability of selecting:

   (a) A ticket marked W?

   Answer _____[2]

   (b) A ticket **not** marked W?

   Answer _____[2]

16. **Sofa....$960**
     Down Payment......$96
     Balance... 12 equal monthly payments

(a) How much is the balance?

   Answer _____[2]

(b) How much is each monthly payment?

   Answer _____[2]

17. A class has 14 boys and 18 girls

(a) Write the ratio of boys to girls in its lowest terms.

   Answer _____[2]

(b) Write the number of girls as a percentage of the class.

   Answer _____[3]

18. (a) Use a pencil, ruler and a pair of compasses to draw a circle of radius 4cm.

[2]

(b) Insert the following on the circle above.

   i.   Centre B, [1]
   ii. Diameter ABY, [1]
   iii. Radius BC [1]

PRACTICE TEST #10

FOR OFFICIAL USE ONLY	
TOTAL MARKS	/50

SCHOOL NUMBER	CANDIDATE NUMBER
INITIAL(S)	SURNAME

INSTRUCTIONS

- Write your school number, candidate number as well as your initial(s) and surname in the spaces provided.
- Answer **ALL** questions in the spaces provided.
- **ALL** working *must* be shown,
- The use of calculators, slide riles, tables or other calculation aids is **NOT** allowed.
- **ALL** working is to be done in **blue** or **black ink**. Working and answers written in pencil, except **constructions and graphs**, may not be marked.
- **ALL** diagrams are not drawn to scale unless otherwise indicated.
- The mark for each question, or part question is shown in brackets [ ]

1.
(a)   4163
    + 875
      21
     ____

    Answer _____[1]

(b)   8756
    - 5490
     ____

    Answer _____[1]

2.
(a)   3159
    x   9
     ____

    Answer _____[1]

(d)   5715 ÷ 3

    Answer _____[1]

3. Round 5 684 to the nearest 10

    Answer _____[1]

4. Write the next number in each of the sequences below

   (a) 4, 9, 16, _____   [1]

   (b) 39, 36, 33, 30, _____   [1]

BJC Mathematics Practice Tests –Paper One

5. Complete the shape about the line of symmetry (dotted line).

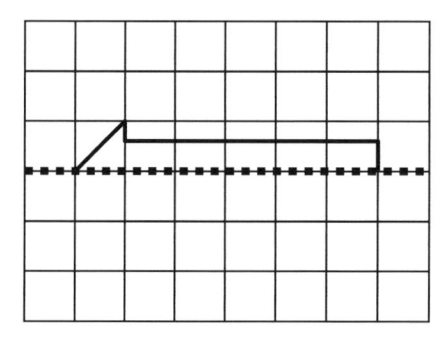

[2]

6. Use your ruler and compass to bisect the line AB below

A _____ B

[2]

7. Given that

90 = 2 × 3 × 3 × 5

54 = 2 × 3 × 3 × 3

Find the Highest common factor (HCF) of the two numbers

Answer _____[2]

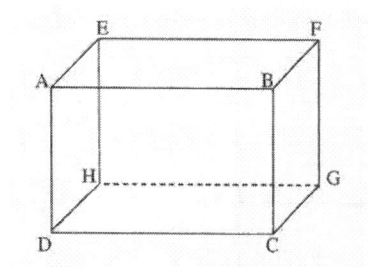

8.

AB is parallel to DC

(a) Name the line parallel to DH

Answer _____[1]

BC is a vertical line

(b) Name the other vertical line

Answer _____[1]

9. Express the fraction $\frac{35}{50}$ as a percent.

Answer _____[2]

10. From the numbers in the table below, write:

51	18	23
27	7	5
12	3	9

(a) The largest prime number.
Answer _____[1]

(b) A cube number
Answer _____[1]

(c) A multiple of 6
Answer _____[1]

11. Pineapple Airline allows its passengers to travel with 1 bag and a carry on at no extra cost. The second bag and all others are at an additional cost of $30 each.

   (a) How much would a passenger with 3 bags and a carryon bag have to pay?

   Answer $_____ [2]

   (b) The first bag weighs 7.85kg. Express this in grams.

   Answer _____ [1]

12. Lataija had 120 candies. She shared 90% among her friends.

   (a) How many candies did she share among her friends?

   Answer _____ candies [2]

   (b) If each friend received 6 candies, how many friends did she share with?

   Answer _____ friends [2]

BJC Mathematics Practice Tests – Paper One

Page 101

13.

30 cm   P = 112 cm   NOT TO SCALE

The perimeter of the rectangle above is 112 cm. The length is 30cm. Calculate the width of the rectangle.

Answer _____[3]

14.

BOOKS – SPECIAL OFFER

Option 1   **Option 2**
$1.49 each   OR   3 for $4.35

(a) Lance bought 3 books at $1.49 each. How much did he pay altogether?

Answer _____[2]

(b) What is the difference in price between using Option 1 and Option 2?

Answer _____[2]

15.

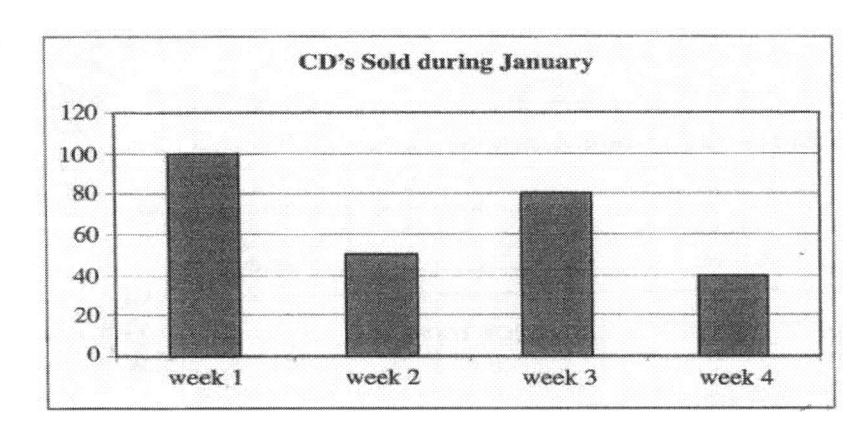

CD's Sold during January

(bar chart: week 1 = 100, week 2 = 50, week 3 = 80, week 4 = 40)

(a) In which week was the largest number of CD's sold?

Answer _____ [1]

(b) In which week was the least number of CD's sold?

Answer _____ [1]

(c) How many more CD's were sold in week 3 than in week 2?

Answer _____ [2]

16. Calculate the value of the lettered angles

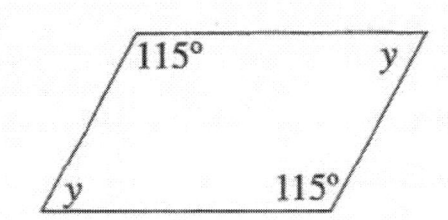

(parallelogram with angles 115°, y, y, 115°)

(a)

Answer: y = _____ [3]

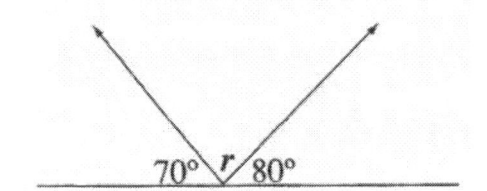

(b)

70° r 80°   NOT TO SCALE

Answer: r = _____ [2]

17. Simplify the following expressions:

(a) $3m + 8m + 6z - 2z$

Answer _____ [2]

(b) If f = 3, g = 5, h = 10 find the value of

$2f - g + h$

Answer _____ [2]

(c) Solve for d

$3d + 6 = 18$

Answer _____ [3]

PRACTICE TEST #11

FOR OFFICIAL USE ONLY	
TOTAL MARKS	/50

SCHOOL NUMBER	CANDIDATE NUMBER
INITIAL(S)	SURNAME

INSTRUCTIONS

- Write your school number, candidate number as well as your initial(s) and surname in the spaces provided.
- Answer **ALL** questions in the spaces provided.
- **ALL** working <u>must</u> be shown,
- The use of calculators, slide riles, tables or other calculation aids is **NOT** allowed.
- **ALL** working is to be done in **blue** or **black ink**. Working and answers written in pencil, except **constructions and graphs**, may not be marked.
- **ALL** diagrams are not drawn to scale unless otherwise indicated.
- The mark for each question, or part question is shown in brackets [ ]

1.
(a)   137
    + 4062
      258
     _____

         Answer _____[1]

(b)   8097
    - 5934
     _____

         Answer _____[1]

2.
(a)   2870
    x   6
     _____

         Answer _____[1]

(b)   1795 ÷ 5

         Answer _____[1]

3. Write the letter of the diagram which represents a pentagon.

   A   B   C   D

         Answer _____[1]

4. Round 3.463 to the nearest hundredth.

Answer _____[1]

5. Draw the line(s) of symmetry for the diagram below.

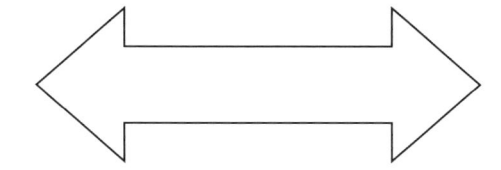

Answer _____[1]

6. Express 0.64 as a fraction in lowest terms.

Answer _____[2]

7. Simplify $20 - 4 \div 2$

Answer _____[2]

8. **CHAIRS - $24.99 each**

    Calculate the total cost of 9 chairs.

    Answer $_____[2]

9. Calculate the size of angle x.

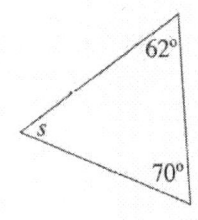

    Answer _____[2]

10. Arrange the integers below, in order of size, from the least to the greatest.

    +7, -7, 0, +3, -3

    Answer _____[2]

11. Evaluate

    (a) $2^5 - 3^2$

    Answer _____ [3]

    (b) $\sqrt{25} + \sqrt{64}$

    Answer _____ [3]

12. Five poles are each placed 110 metres apart.

    (a) How far will Trevor travel when he walks from the second pole to the fifth pole?

    Answer _____ m [2]

    (b) Express your answer in (a) as **kilometers**.

    Answer _____ km [2]

13. A call from Mangrove Cay to Conch Creek cost $1.50 for the first minute then 35¢ for each additional minute. What is the total cost of a call from Mangrove Cay to Conch Creek which lasts 5 minutes?

Answer $ _____ [3]

14. From the set of numbers {51, 7, 11, 46, 93} write down

   (a) One composite number

   Answer _____ [1]

   (b) One even number

   Answer _____ [1]

   (c) One multiple of 23

   Answer _____ [1]

15. Three brothers, Ackeem, Brandon and James were given $264 to share in the ratio 2 : 3 : 1 respectively.

How much money did James receive?

Answer $ _____ [2]

16.

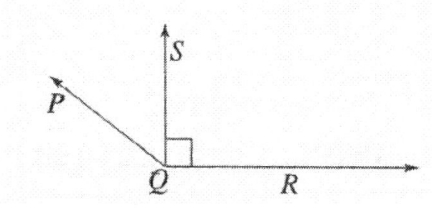

Use the figure above to answer the questions below.

(a) _____ is an acute angle. [1]

(b) _____ is an obtuse angle. [1]

(c) _____ is a line perpendicular to another line. [1]

17. A Bible Club had 160 members. There was a membership increase of 5%.

(a) How many new persons joined the club?

Answer _____ [2]

(b) How many persons are now members of the Club?

Answer _____ [2]

18. Then Venn diagram below shows the results of a survey of a class.

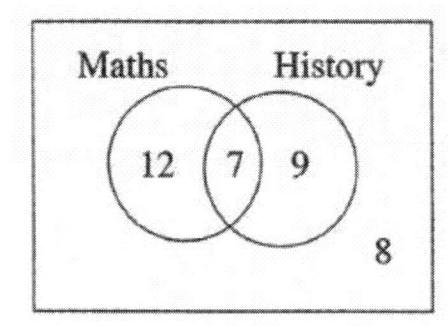

Maths: 12, 7 (intersection), History: 9, Outside: 8

(a) How many students are in the class?

Answer _____ [1]

(b) How many students take neither History nor Maths?

Answer _____ [1]

(c) How many students took both subjects?

Answer _____ [1]

(d) How many students took Mathematics only?

Answer _____ [1]

19. (a) Solve for x

$$\frac{x}{5} = 7$$

Answer _____[2]

(b) If $f = 4$, $g = 5$ find the value of

$$g + f - 6$$

Answer _____[2]

PRACTICE TEST # 12

FOR OFFICIAL USE ONLY	
TOTAL MARKS	/50

SCHOOL NUMBER	CANDIDATE NUMBER
INITIAL(S)	SURNAME

INSTRUCTIONS

- Write your school number, candidate number as well as your initial(s) and surname in the spaces provided.
- Answer **ALL** questions in the spaces provided.
- **ALL** working _must_ be shown,
- The use of calculators, slide riles, tables or other calculation aids is **NOT** allowed.
- **ALL** working is to be done in **blue** or **black ink**. Working and answers written in pencil, except **constructions and graphs**, may not be marked.
- **ALL** diagrams are not drawn to scale unless otherwise indicated.
- The mark for each question, or part question is shown in brackets [ ]

1.
(a) 305
    + 72
    2469
    ─────

    Answer _____ [1]

(b) 9024
    - 6173
    ─────

    Answer _____ [1]

2.
(a) 6507
    x   3
    ─────

    Answer _____ [1]

(d) 8406 ÷ 6

    Answer _____ [1]

3. Write down the value of:

   8 hundreds, 3 tens and 6 ones

   Answer _____ [1]

4. Express 0.306 correct to the nearest hundredth.

   Answer _____ [1]

5. What fraction of the shape is **UNSHADED**?

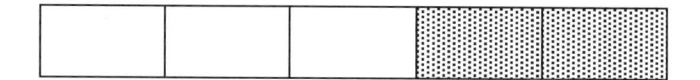

Answer _____[1]

6. Write the letter that is beneath the pair of lines which best represent parallel lines.

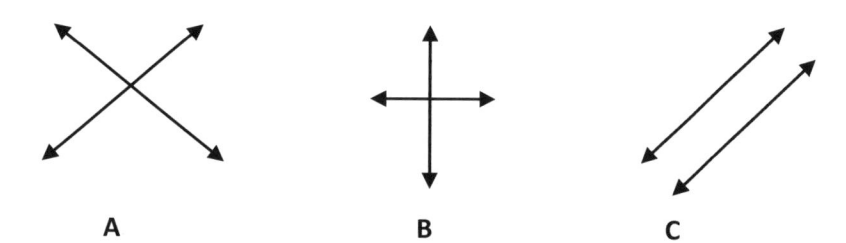

A   B   C

Answer _____[1]

7. Given that 1 cm represents 50 m, find the actual distance between the lighthouse and the boat (AB).

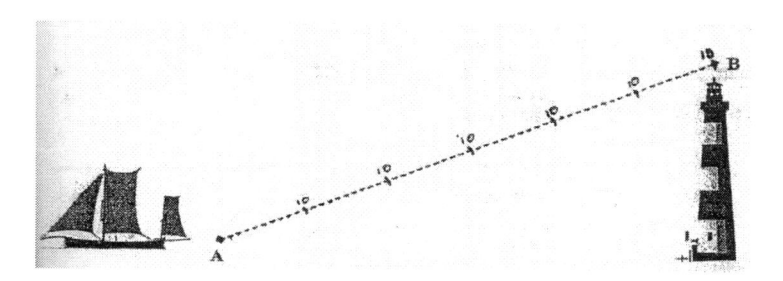

Answer _____[2]

BJC Mathematics Practice Tests –Paper One

8. The cost for a cell phone call with LIVE's cell phone card is 40 cents per minute. How many minutes can I buy for $20?

Answer _____ mins [2]

9. Use the diagram below to answer the questions which follow.

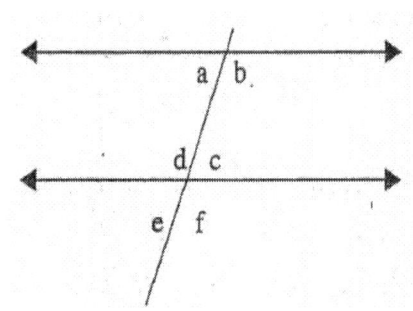

(a) Identify a pair of corresponding angles.

Answer _____ [1]

(b) Angle a = 62°. Write down the size of angle c.

Answer _____ [1]

10. Draw in, where possible, the line(s) of symmetry on the letters below.

DARE

[3]

11. Glass A has 83 *ml* of liquid and glass B has 125 *ml* of liquid.

   (a) What is the total amount of liquid in both glasses?

   Answer _____ml [1]

   (b) Express your answer to (a) in litres.

   Answer _____litres [2]

12. Write six and four-ninths as an improper fraction.

Answer _____[3]

13.   Omar's Taxi Service

1ˢᵗ Mile..............................$5.50
Each additional mile................$2.25

Omar's Taxi Service charges $5.50 for the first mile and $2.25 for each additional mile. How much will he charge for a journey of 9 miles?

Answer _____[3]

14. Use your pencil, ruler and protractor to draw X Ŷ Z= 70°

[3]

15. Multiply $1\frac{3}{4}$ by $2\frac{1}{5}$

Answer _____ [4]

16.

Breanna	Shekinah	Miracle
2	3	4

Breanna, Shekinah and Miracle shared $846 in the ratio 2:3:4 respectively. How much money did Miracle receive?

Answer _____ [5]

17. The students at Skills Tutor School held a canned food drive. The results are shown below.

GRADE	CANS COLLECTED
7	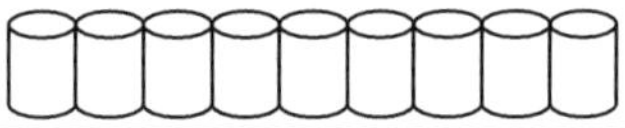 🥫🥫🥫🥫🥫🥫🥫
8	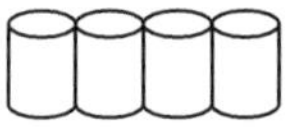 🥫🥫🥫
9	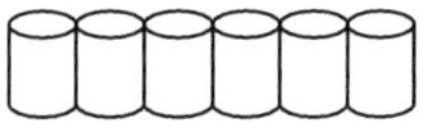 🥫🥫🥫🥫🥫
10	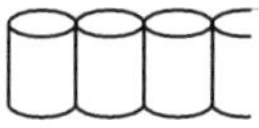 🥫🥫🥫½
11	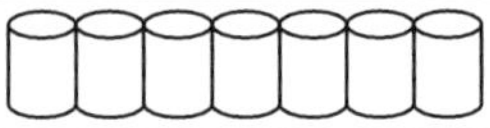 🥫🥫🥫🥫🥫🥫
KEY:	 🥫 represents 20 cans

(a) Which grade collected 70 cans?

Answer _____[1]

(b) How many more cans did **Grade 7** collect than **Grade 11**?

Answer _____[3]

(c) How many cans were collected in all?

Answer _____[2]

BJC Mathematics Practice Tests – Paper One

18. (a) Simplify

   $6m + p + 3p - 2m$

   Answer _____ [2]

   (b) Given that $p = 7$, calculate the value of $2p^2$

   Answer _____ [2]

   (c) Solve for 'h'

   $3h - 5 = 19$

   Answer _____ [2]

BJC Mathematics Practice Tests – Paper One

Made in the USA
Columbia, SC
27 October 2022